Copyright Danny Spinks, 2014

Dedications

This book is dedicated to My fiancé Chung Mun Noah, who has been instrumental in her support, help and encouragement. Her poetry has been inspirational, she has helped motivated me to complete this book project. I've always been interested in Earth Science. As a child my father would often joke with me about my interest in the solar system by saying "the moon is made of cheese ".

As a child I knew it wasn't but it begged the question: what was the moon really made of? My father Oliver Hypes Spinks and mother Violet Marie Spinks encouraged my interest in science. I developed a curiosity about the earth and the moon.

I would like to thank my mother and father for encouraging my interest in earth science.

I would also like to thank Cooper Benedict and Pickney Benedict for their stimulating conversations about earth and Space.

A special thank you goes to both:

Cora Glickstein, Editor and

Bill Hastings, Editor and book production Manager.

Preface:

Spinks Books: Easy to Read, Easy to Teach from series:

Spinks books are designed for children and young adults.

I envision this book being used by elementary and middle school classes as an introduction to Earth Science.

Spink Books are a series of Math, Science and History, E books and paper back books.

These books are intended to be a fun, easy to read and understand, for children in elementary and middle schools.

Easy to teach from!

This series is designed with the teacher in mind. Each Spinks book is an **Easy Introduction** to the subject, written in simple

My mission is to motivate children to learn, get excited about a subject and be very easy for the teacher to teach from.

From my experience teaching developmentally challenged children, I know that every child can be motivated to learn and achieve his potential.

I have learned the value of easy to use teaching materials and the difficulty finding those materials.

Spinks books are intended to fill a void of easy to read and use teaching materials and books.

Danny Spinks, Author, Spinks Books.

CHAPTER 1

EARTH

To us, our planet Earth seems enormous, but if we were able to gaze at it across the vastness of space, it would look like a tiny speck. It is one of the nine planets that are constantly hurtling around a Star – our Sun – along individual oval paths, called orbits.

Together the Sun and its planets are known as the Solar System. This, in turn, is part of a cluster of millions of stars and planets, called a galaxy. Our galaxy, which is shaped like a spiral, is called the Milky Way. It is so huge that a jet would take 100 trillion years to fly across it. So far, we have discovered about 6 billion different galaxies, and together they make up the universe.

Stars are made from layers of burning gas around a dense core. Some planets are mostly gas, too, while others are liquid. Our Earth, however, is like a gigantic ball of rock.

The Earth Space (1)

THE EARTH

Imagine you are an astronaut looking at the Earth from your spacecraft. What you see is a big blue ball covered with swirling clouds that hide features such as continents and mountains. The ball looks so blue because more than two-thirds of it is covered with water in the form of oceans, seas, lakes, and rivers.

The surface of the Earth, called the crust, moves all the time, but this movement is so slow that we are not aware of it. Eventually, however, pressure builds up and causes earthquakes and volcanoes. Changes also happen when the crust is worn away by water or huge blocks of moving ice, called glaciers.

Our planet can support life only because it gets light and heat from the Sun. Without it, the Earth would be a cold, dark, and dead place. (2)

The Himalaya Mountains in Nepal look very different from far out in space. (3)

Stalactites form in caves. (4a)

Sandstone (4)

Pumice (5)

Green Marble (6)

Ayers Rock in Australia (7)

Earth photographed from a satellite (8)

THE EARTH'S CRUST

Just like you, the Earth has a very delicate skin. It is so thin that if you compare it to the whole Earth, it is thinner than the skin of an apple.

The Earth's skin, or crust, is made up of rock, built up in layers over millions of years. The layers look like blankets on a bed, with lots of lumps and bumps in them.

The crust is a thin layer of rock between 3-1/2 and 42 miles (5.6 and 67.6 km) thick. (9)

The mantle is the layer below the crust. In parts of it, the rock has melted like butter. (10)

The outer core is made of iron and nickel that have melted to form a liquid (11)

The inner core is a ball of iron and nickel. It is hotter here than at the outer core but the ball stays solid. (12)

The ocean is on top of the oceanic crust. The oceanic crust also runs underneath the continental crust. (13)

Under the oceans, the crust is as little as 3-1/2 miles (5.6 km) thick, but under the continents, it is up to 42 miles (67.6 km) thick. (14)

Land is made out of the continental crust. It is thickest where mountains are found. (15)

The mantle (16)

How Mountains Are Made

Mountains are made when the Earth's crust is pushed up in big folds or forced up or down in blocks. The different shapes made are given different names.

(17)

Upfold

Downfold

Overfold

Rift valley

Fault

Block mountain

A Long Way to Go

Did you know that the deepest hole ever drilled into the Earth's crust is 7.8 miles (12.5 km) deep? To reach the center of the Earth, you would have to drill 500 times deeper. (17a)

Going Down

This is a rift valley. It was made when a block of land sank down between two long breaks, called faults, in the Earth's crust. (18)

Going Up

Here the land has been pushed into giant folds by movements in the Earth's crust. You can see how the crust is made up of many, many layers of rock. (18a)

MOVING PLATES

The Earth's crust is not one unbroken piece. It is made up of many pieces that fit together like a giant jigsaw puzzle.

These pieces, called plates, ride on soft, partly melted rock moving underneath them. The pieces push against each other, creating spectacular effects: earthquakes split the crust, volcanoes form, new land is made, and huge mountain ranges are pushed skyward.

On the Move

The plates move all the time. In one year, they will usually move about one inch (2.5 cm). That's about as much as your fingernails grow in the same amount of time.

All Scrunched Up

Sometimes, two plates push against each other and crumple the land, making huge mountain ranges. (19)

Going Down

Sometimes, one plate slides under another. It is pushed down into the mantle and melts. (20)

Doing a Split

Sometimes, two plates split apart, and lava bubbles up to fill the gap. It hardens and forms new land. (21)

Slip-Sliding Away

Sometime, two plates slip sideways past each other. This kind of movement causes earthquakes. (22)

The red dots show you the places where volcanoes erupt.

Continent

The green dots show you the places where earthquakes occur.

This is where two plates meet. (23)

Changing Places

The land is coming together to make one gigantic continent. (24)

All together

The super continent has come together. We call it Pangaea. (25)

Worlds Apart

The land is drifting apart again. Pangaea is splitting into two parts: Laurasia and Gondwanaland. (26)

Familiar Ground

Today, the world looks like this - but the continents are still moving. (27)

Looking Ahead

This is how the world may look 50 million years from now. Can you spot how the land has changed its shape? To get started, find Africa on the globe and see how it has joined up with Europe. (28)

CHAPTER 2

VOLCANOES

When you shake up a can of soda and then pull off the tab, the contents shoot out with a whoosh! A volcano acts a bit like this. With tremendous force, melted rock bursts through weak parts in the Earth's crust and is hurled high into the sky.

Nature's Fireworks

This volcano is putting on its own spectacular fireworks display. The explosions of red-hot lava and ash from the crater look like gigantic Roman candles. (29)

The Spotter's Guide to Volcano Shapes

Spreading Out

Some volcanoes are flat. Their lava is very runny, so it spreads out in a thin sheet. (30)

Short and Plump

Some volcanoes are squat. They are made of ash, which is lava that has turned to dust. (31)

Going Up

Some volcanoes have pointed cones. Their lava is thick and sticky, so it does not run far. (32)

Hot springs are often found near volcanoes.

Volcanoes can be quiet and not erupt for a long time.

Clouds of ash and gas pour out from the crater.

Molten rock, called magma, rises up the main pipe and any branch pipes.

A volcano builds up from layers of ash and lava.

Branch pipe

Magma collects in a chamber found deep underground. It is forced up through cracks and holes in the ground. (34)

Mixed Bunch

When lava cools and hardens, it can make rocks with different shapes. Here are three types:

Runny lava (36)

Ropy lava (37)

A volcanic "bomb" (38)

CHAPTER 3

EARTHQUAKES

Our planet is a restless place. Every 30 seconds, the ground suddenly rumbles and trembles. Most of the movements are so slight that they are not felt. Others bring complete disaster. Big cracks appear in the land, streets buckle, and buildings crumble. Whole towns and cities can be destroyed. Then everything settles down but is totally changed. The Earth has shaken. An earthquake has happened. (40)

Unsafe Ground

This is the San Andreas Fault in California. Earthquakes regularly happen here. (41)

Cars are smashed and they settle at crazy angles.

Fallen telephone lines.

Fires are started by broken gas pipes and broken electrical cables.

Fault line

On this side of the fault the land has moved away from you. (42)

Terror from the Sea

Earthquakes under the ocean can cause giant, destructive waves called tsunamis, or tidal waves.

An earthquake occurs along a fault in the seabed.

Tsunamis can travel many miles across the ocean.

A tsunami can be almost 200 feet (61 m) high and can travel as fast as a jet. (43)

Why Earthquakes Happen

You may think that your feet are firmly on the ground, but the Earth's crust is moving all the time. It is made of moving parts called plates. When the plates slide past or into each other, the rocks jolt and send out shock waves.

Shaken Up

The Mercalli Scale measures how much the surface of the Earth shakes during an earthquake. There are 12 intensities, or grades. At intensity 1, the effects are not felt, but by intensity 12, the shock waves can be seen and there is total destruction. (44)

What to Do in an Earthquake

Indoors, lie down under a bed or heavy table, or stand in a doorway or a corner of a room. After a minute, when the tremors will usually have finished, go outside, away from buildings, to a wide-open space. (45)

Earthquake Words

The place within the Earth where an earthquake starts is called the focus. The earthquake is usually strongest at the epicenter. This is the point on the Earth's surface directly above the focus. The study of earthquakes and the shock waves they send out is called seismology. (46)

Destructive Force

A tsunami piles up and gets very tall before it crashes onto the shore. It is so powerful that it can smash harbors and towns and sweep ships inland.

CHAPTER 4

ROCKS

Movements in the Earth's crust are slowly changing the rocks that make up the surface of our planet. Mountains are pushed up and weathered away. The fragments are moved and made into other rocks. These rocks may be dragged down into the mantle and melted by its fierce heat. When a volcano erupts, the melted rock is thrown to the surface as lava, which cools and hardens as rock. This is broken down by weathering, and so the cycle starts again. (47)

In the Beginning

Rocks belong to three basic types. Igneous rocks are made from magma or lava and are also known as "fiery" rocks. Sedimentary rocks are made in layers from broken rocks. Metamorphic rocks can start off as any type. They are changed by heat and weight and are called "changed form" rocks.

Sedimentary Rocks

These are made from bits of rock and plant and animal remains. They are broken into fine pieces and carried by rivers into the sea. They pile up in layers and press together to make solid rock. The Painted Desert in Arizona is made of sedimentary rocks. (48)

Conglomerate (49)

Limestone (50)

Red sandstone (51)

Obsidian (52)

Granite (53)

Igneous Rocks

These are made from magma or lava. It cools and hardens inside the Earth's crust or on the surface when it erupts from a volcano. Sugar Loaf Mountain in Brazil was one igneous rock under the crust. The rocks above and around it have been worn away.

Metamorphic Rocks

These are igneous or sedimentary rocks that are changed by underground heat, underground weight, or both. This marble was once limestone, a sedimentary rock. It was changed into marble by intense heat.

Marble (54)

Slate (55)

In time, material moved by rivers and piled up in the ocean will become sedimentary rocks.

Rock fragments are carried from one place to another by rivers, glaciers, the wind, and the sun.

Surface rocks are broken down by the weather and by the scraping effect of tiny pieces of rock carried in the wind or in the ice of glaciers.

Glacier

Some rocks are thrust up as mountain ranges when the crust moves and makes giant folds.

Volcano

CHAPTER 5

CAVES

Caves are hollows beneath the surface of the Earth. The biggest ones are all found in rock called limestone, and some are huge. The world's biggest cave, in Sarawak in Borneo, is so large it could fit 800 tennis courts in it. Yet these caves began simply as cracks or holes in the rock that, over thousands of years, were made bigger by rainwater trickling into them and eating them away.

Drip . . .

The rainwater that seeps into the ground is very slightly acid and begins to eat away the limestone. (57)

Drip . . .

The rainwater eats through the rock. It widens the cracks into pits, passages, and caves. (58)

27

Drip

Over thousands of years, the passages and caves may join up to make a huge underground system. (59)

Going Down

Water dripping from the ceiling of a cave leaves behind a mineral called calcite. Very slowly, this grows downward in an icicle shape that is called a stalactite. (60)

Limestone is a very common rock. It is made from the skeletons and shells of tiny sea creatures that died millions of years ago.

This pothole, or tunnel, leads straight down through the rock. It was made by a stream wearing away the rock.

The stream disappears underground into a pothole.

Cracks in the rock are widened when rainwater seeps along them.

Limestone pavements are made when the rock is eaten away along joint lines.

Cliff

Gallery

Cave mouth

Stream (61)

Tunnel of Lava

Caves are found in rocks other than limestone. This one is made of lava and is inside a volcano in Hawaii. (62)

Caving In

Sometimes a cave turns into a gorge. This happens when the roof falls in to reveal underground caverns and, far below, the river that carved them. (63)

Going Up

Where water drips onto the cave floor, columns of calcite, called stalagmites, grow upward. (64)

CHAPTER 6

OCEANS

More than two thirds of our planet is covered with water, and oceans make up 71 percent of the Earth's surface. Beneath the seas lies a fascinating landscape. Much of the ocean floor is a vast plain, but there are also cliffs, trenches, and mountain ranges, all larger than any found on dry land. (65) (66)

Ocean Currents

These show the directions in which water flows.

Cold currents (67)

Warm currents (68)

Ebbing and Flowing

Tides are made by the Sun and Moon pulling on the oceans. When the Sun, Earth, and Moon are in a line, there are large spring tides. (70)

Underwater canyons are cut by currents flowing over the seabed like rivers.

These underwater islands are called guyots.

Trenches can be deeper than the highest mountains on land.

This island is a volcano that has erupted from the ocean floor.

A long, wide ocean ridge (71)

Going . . .

The water inside a wave moves around and around in a circle. It is the wind that drives the wave forward. (72)

Going . . .

Near the shore, the circular shape of the wave is changed, and it becomes squashed. (73)

Gone

The top of the wave becomes unstable. When it hits the beach, it topples and spills over. (74)

Ocean Currents

The direction in which currents move depends on winds and the Earth's spin. Winds blow the top of the oceans forward, but the Earth's spin makes the water below go in a spiral. (75)

The Dark Depths

Even in clear water, sunlight cannot reach very far. The oceans become darker and darker the farther down you go - until everything turns inky black.

Surface

Lighted zone 660 feet (200 m)

Dark zone 20,000 feet (6,096 m)

Deepest zone, a trench of 36,300 feet (11,064 m) (76)

Frozen Worlds

In Antarctica and the Arctic, the oceans freeze. Icebergs break away from glaciers flowing into the water. Only a tiny part of an iceberg is seen above the surface of the ocean. (77)

CHAPTER 7

COASTLINES

Have you ever built a sand castle and then watched the sea come in, knock it down and flatten it? This is what happens to the coastline, the place where the land and the sea meet. The coastline changes all the time because, every few seconds of every day, waves hit the land and either wear it away or build it up into different shapes.

Going, Going, Gone

When caves made on both sides of a headland meet, an arch is formed. If the top of the arch falls down, a pillar of rock, called a stack, is left.

An arch

Headland

A stack

A cave forms over thousands of years as seawater creates cracks and holes in a cliff and makes them bigger.

Some beaches are made in bays between headlands where the water is shallow and the waves are weak.

Some waves carry sand and pebbles from one area of the coast and leave them at another. This forms a new beach.

Mud flats and marshes

An estuary is the place where a river flows into the sea.

Waves can build sand, mud, and pebbles into a long strip of new land, called a spit.

Sea cliffs are one of the best places to see the different layers of rock. (78)

Pounding Away

Waves pound the coastline like a giant hammer until huge chunks of rock are broken off. The chunks are then carried away by the sea and flung against the coastline somewhere else. (79)

Shifting Sands

Dunes are made of sand blown into low hills by the wind. (80)

From Rocks to Sand

Waves roll rocks and boulders backward and forward on the shore. The boulders break into pebbles and then into tiny grains of sand. This change takes hundreds or thousands of years. (81)

Living Rock

Coral is found in warm, sunny, shallow seas. It is made by tiny sea creatures that look like flowers. Over thousands of years, their skeletons build up into huge coral reefs and islands. (82) (83)

CHAPTER 8

GLACIERS

A glacier is a huge river of ice that starts its life as a tiny snowflake. As more and more snow falls and builds up, in time it gets squashed under its own weight and turns to ice. A glacier moves very slowly downhill. Because it is very heavy, it can push rock along like a bulldozer. It can wear away the sides of mountains, smooth off the jagged bits from rocks, and move giant boulders over tens of miles. (84)

Close-Up View

The pilot in this plane is watching a wall of ice break away from a glacier and begin to crash into the water below. (85)

Mountains

Glaciers begin as huge snowfields

The snow collects in hollows and turns to ice under its own weight.

Glaciers usually move downhill very slowly - no more than an inch or two (2.5 or 5 cm) a day.

The ice begins to move and rub away the sides and bottom of the hollow. Little by little, it changes the shape of the land and makes it into a U-shaped valley.

Rubble is carried along by the glacier.

Melted ice flows as streams and rivers inside most glaciers. (86)

Ice Power

When the water in this bottle freezes and turns to ice, it takes up more room and breaks the bottle. When the water that makes up the ice of a glacier freezes, it takes up more room and pushes away the rock. (87)

Bumps in the rock can be smoothed out by the ice moving downhill.

Rocks carried along by the glacier pile up when the glacier starts to melt and stops pushing them.

The lower end of the glacier is called the "snout."

When the glacier melts, it makes new rivers.

Where a glacier flows into water, chunks of ice break off and float away. (88)

Shaping the Land

When you see a valley like this, you can tell from its U shape that it was once filled with the ice of a glacier. (89)

Out of Place

This giant boulder of hard rock was moved by a glacier and left on soft limestone. Then most of the limestone was weathered away, leaving a small block under the boulder. (90)

CHAPTER 9

RIVERS

Rivers are very powerful, so powerful that the force of the moving water is able to change the shape of the land. As they flow through mountains and over plains, rivers carry away huge amounts of rock, sand, and mud. They then dump it somewhere else, usually on riverbanks or in the ocean, to make new land.

Over the Top

When a river tumbles over the edge of a steep cliff or over a hard, rocky ledge, it is called a waterfall. This one is in Brazil. (91)

As the river flows quickly down steep slopes, it wears away the rock to make a V-shaped valley.

Sand, mud, and gravel are left by the water as sediment.

Glacier

A river usually begins in mountains or hills. Its water comes from rain or melted snow.

Where the rock is hard, the river makes rapids or waterfalls.

A wide bend or meander goes across flat country.

As it reaches the sea, the river divides into small streams, leaving a mass of sand, mud, and rock fragments, called a delta. (92)

River Deep

A deep, narrow valley carved by a river is called a gorge. The Colorado River in the U.S.A. has made the world's largest gorge - the Grand Canyon. It is 1 mile (1.6 km) deep and 218 miles (350 km) long. (93)

River's End

This swampy land is part of the Yukon Delta in Alaska. (94)

Around the Bend

When a river reaches flat land, it slows down and begins to flow in large loops. It leaves behind sand, gravel, and mud, called deposits. This changes the river's shape and course. The river leaves deposits on the inside bend and eats away the outer bend. (95)

The deposits change the shape of the bend. In time, the neck of the bend narrows and the ends of the neck join up. (96) The river leaves behind a loop. It is called an oxbow lake because of its shape. (97)

Record Rivers The longest river in the world is the Nile, in Africa. It is 4,160 miles (6,690 km) long. The largest delta covers 30,000 square miles (77,700 km). It is made by the Ganges and rahmaputra Rivers, in Bangladesh and India.

CHAPTER 10

DESERTS

Did you know that deserts come in many different forms? They can be a sea of rolling sand, a huge area of flat and stony ground, or mountainous area of shattered rock. There are hot deserts and cold deserts. So what do these very different areas have in common?

The answer is that they are all very dry, and they all get less than 10 inches (25 cm) of rain each year. This rain may not fall regularly. Instead, it may all come in a single day and can cause a dramatic flash flood.

Wind Power

Wind blows the sand into hills, which are called dunes. These have different shapes and names.

Tail dune (98)

Barchan dunes (99)

Seif dunes (100)

Star dunes (101)

Dunes

Butte

Chimney or pipe rock

Natural rock arch

Cuesta

Mesa

Heave rain causes flash floods. These rush over the land, loaded with sand

and stones, and cut deep channels in the surface of the desert.

Broken rocks slide downhill and collect in gullies.

Where the rock is hard, ridges will stand out in the landscape.

Outwash fan

Steep slopes of broken rock (102)

Sea of Sand

A desert may be hard to live in, but it can be stunning to look at. These dunes are in Saudi Arabia. (103)

On the Move

Imagine your hair dryer is the wind. It blows sand up the gentle slope of the dune. When the sand goes to the top, it tumbles down the steep slope. As more and more sand is moved from one slope to another, the whole dune moves forward. (104)

Water Power

The tremendous power of water has made this deep ravine near an oasis in Tunisia. (105)

Shaping the Land

Wind-borne sand blows against the rocks and wears them into beautiful and surprising shapes. (106)

Hot and Cold

This is one of the Devil's Marbles in the Northern Territory, Australia. The rock's outer layers have started to peel off because of the desert's very hot and very cold temperatures. (107)

CHAPTER 10

SPACE

The universe is made up of galaxies, stars, planets, moons, and other bodies scattered throughout space. A galaxy is a group of millions of stars. Our galaxy, which is shaped like a spiral, is called the Milky Way.

On a clear night, it is possible to see thousands of stars, which appear as twinkling points of light. Earth's moon is usually clear, and sometimes you can also see five of the Planets: Mercury, Venus, Mars, Jupiter, and Saturn. These do not twinkle. They look like small, steady disks of light. Earth is the third planet from the Sun, which is about 93 million miles (149 million kilometers) away from us.

People have always been curious about the things they could see in the sky. It is only quite recently, though, that science has developed the advanced technology needed to send men and women into space. (108)

FIG. 108

Pluto, Neptune, Uranus, Saturn, Rings, Jupiter, Mars, Earth

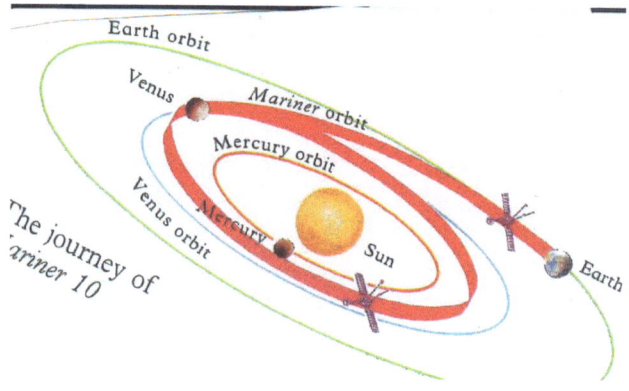

Fig 108 The Flight path of Mariner 10 space craft in red exploring the planets!

ROCKETS

Rockets were invented in China a long time ago. They looked a bit like arrows and worked by burning gunpowder that burned up very quickly, so the rockets did not travel very far. Since then, people have tried many ways of sending rockets up into space. In modern rockets, two liquid fuels are used. They mix together and burn. Then the hot gas shoots out of the tail, pushing the rocket

up and away.

See It Go!

If you blow up a balloon and let it go without tying a knot in the neck, the air will rush out very quickly. When the air goes out one way, it pushes the balloon the other way - just like a rocket! (114)

3,2,1, Fire!

A hundred years ago, soldiers used rockets like this. They were called Congreve rockets. (115)

The Fly!

In 1931, a German named Johannes Winkler launched his HW-1 rocket. It went 7 feet (2 m) into the air, tuned over, and fell back to the ground. A month later, he tried again. This time it climbed to 297 feet (90 m) and landed 660 feet (201 m) away. (116)

V-2 rocket

   1945 (117)

Gemini Titan

   1964 (118)

Soyuz

   1967 (119)

Space shuttle

   1981 (120)

Saturn Power

Saturn 5 is the biggest rocket ever built. It is as tall as a 30-story building! It was used in the Apollo program, which carried the first American astronauts to the Moon.

The stabilizing fins keep the rocket on course. (121)

Five rocket engines (122)

Fuel tank (123)

Five rocket engines (124)

Rocket engine (125)

Lunar module (126)

Service module (127)

Command module (128)

Launch escape system (129)

Overpowering

See just how enormous Saturn's engines are compared with these people! (130)

Sky High

The space shuttle leaves the launch pad in a blaze of bright light. It will circle the Earth over a hundred times in about eight days. Its wings then allow it to glide back to the ground so it can be used again. (131)

Up, Up,......and Away

How far can you throw a ball? About 50 or 60 feet (15 or 18 m)? It doesn't go on forever because the Earth's gravity pulls it back down again. (132)

MOON MISSION

The Moon is the Earth's nearest neighbor in space, but it sill takes three days to get there by rocket. It would take 200days by car! When astronauts first went to the Moon, no one knew if it would be safe to land there. But American astronauts have been to the Moon on Apollo missions six times, and they all returned safely to Earth. The first Moon trip was in 1969 and the last in 1972.

We Have Liftoff!

The first stage of the Saturn 5 rocket has five huge engines. When these run out of fuel, they fall back to Earth. Then the second stage takes over. (133)

Second stage (134)

The second stage drops off when its five engines run out of fuel. (135)

The command and service modules turn, join on to the lunar module, and pull it out of the third stage. (136)

The lunar module drops down to the Moon with two astronauts inside. The command module stays in orbit around the Moon. (137)

The top part of the lunar module returns to the command module. (138)

The command module falls to Earth, using parachutes, to make a safe landing in the sea. (139)

Splashdown!

The command module falls through the Earth's atmosphere so quickly that the bottom is burned. It splashes down in the ocean and is picked up by a helicopter. The balls on the roof are air balloons, which help it float upright if it turns over in the water. (140)

Lunar Module (141)

Command Module (142)

Service Module (143)

This air recycling unit keeps the air fresh in the cabin. (144)

Astronauts traveling to the Moon crawl through a tunnel from the command module to the lunar module. (145)

There is not much room in the cabin for three astronauts. (146)

Control panel (147)

Fuel tanks (148)

Engine nozzle (149)

Hot soda, acting like rocket fuel.

Moon Trail

The Apollo mission to the Moon followed a path in the shape of the figure eight. (150)

LUNAR LANDING

A lunar landing is a Moon landing. If yiou went to the Moon, you would find no living things at all, no air, and no water. If you stayed for a lunar "day" - about 28 Earth days - you would have two weeks of baking sun followed by two weeks of freezing nights. The first people on the Moon went down in the lunar module known as *Eagle*.

Hanging Out the Wash?

No, just setting up a panel to collect dust! The Moon is covered in dusty soil and scattered rocks. (151)

The Apollo 11 Crew

Neil Armstrong and Edwin "Buzz" Aldrin were the first people to walk on the Moon. Michael Collins stayed in orbit in the command module. (152)

Earthrise

The lunar module is just leaving the Moon. Behind it, you can see what the Earth looks like from the Moon. (153)

Ladder

Landing foot with rounded footpad to keep the leg from sinking into the soft dust. (154)

One of the four fold-up "spider" legs. The first lunar module was nicknamed "Spider", but that mission didn't land on the Moon. (155)

*Forward hatch door (*156)

These engines help the astronauts control the lunar module. (157)

Weather or Not?

With no wind or rain, the footprints made by the astronauts will remain on the Moon forever. The American flag, left on the Moon by the astronauts, is held out by a metal bar because there is no wind to make it fly. (158)

Moon Buggy

This open car was taken to the Moon for the first time in Apollo 15. Its correct name is the lunar roving vehicle.

Seats (159)

Television camera (160)

Antenna (161)

Control panel (162)

Hand control (163)

Sample collection bags (164)

Wire-mesh wheel (165)

Space for storing equipment (166)

Super Saturn

The Saturn 5 has three stages. When one stage runs out of fuel, it falls off and another part takes over.

First stage (167)

Second stage (168)

Third stage (169)

Lunar   module   (170)

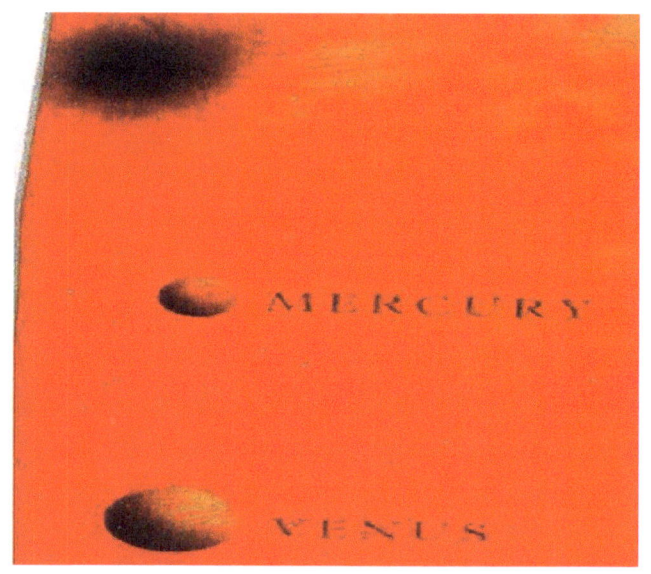

Service module (171) Command module (172),   Escape tower (173)

MERCUTY AND VENUS

Between the Earth and the Sun are two planets called Mercury and Venus. They are very hot because they are the Sun's nearest neighbors. Venus is the brightest object in the night sky, while Mercury is the second smallest planet. Photographs from space probes tell us about these planets.

Hello, Goodbye!

Mariner 10, intended to explore Mercury and Venus, was the first probe to visit two planets in turn. It worked for 17 months before breaking down. It is now in orbit around the Sun,

The journey of Mariner 10 (174)

The solar detector made sure that the solar panels were always facing the Sun. (175)

Television cameras sent pictures back to Earth. (176)

Antenna (177)

Solar panel (178)

Star detector (179)

Mercury (180)

*Crust (*181)

Hard Center

If you could slice Mercury a you would

a peach, you would find a core made of

iron.

Iron core (182)

Mariner (183)

Magellan 184)

Wish you were here?

The Magellan probe used solar cameras to take pictures through the thick fog around Venus. Computers made this 3D image of the volcanoes. (185)

VENUS (186)

Hot Orange

Venus is the hottest planet of all - so hot that it could melt lead! It has a bright orange sky with flashes of lightning. The Earth spins around once every 24 hours, but Venus spins very slowly - once every 244 Earth days!

Venus Venera

Several Russian missions have been to Venus. This program was called

Venera. The spacecraft sent pictures back to Earth. This is the part of Venera 9 that went down to Venus by parachute. (187)

The landing ring helped make the landing soft. (188)

Instrument container (189)

The brake is shaped like a disk to help slow the space probe down. (190)

Venera 9 Venus landing (191)

The space probe was in a capsule on the Venera spacecraft. (192)

The capsule fell through the atmosphere of Venus. (193)

The heat-shield covers separated and fell off. (194)

The probe was slowed down by a small parachute. (195)

Three larger parachutes were used for the final stage. (196)

After a safe landing, the television cameras and instruments were switched on. (197)

THE RED PLANET

Mars is called the red planet because the soil and rocks are red. Light winds blow dust around, which makes the sky look pink. People once thought there was life on Mars, but nothing living has been found so far.

The Viking spacecraft were sent to Mars to find out what it is like. Two missions, Viking 1 and Viking 2, made the journey. Perhaps one day people may go to live on Mars because it is the planet most like our own.

Tight Fit

The Viking lander fits into a capsule on the spacecraft. With its legs folded up, it looks a bit like a tortoise inside its shell. (198)

Viking spacecraft (199)

The Viking lander folded into a capsule on the spacecraft. (200)

It leaves the orbiter and begins its journey down to Mars. (201)

The television camera takes a series of pictures as it moves around. (202)

It moves so fast that it gets very hot. (203)

A parachute is used to slow it down. Then the heat shield drops off. (204)

The legs unfold, and the rockets are used as brakes for a soft landing. (205)

The landing feet have rounded pads to keep the legs from sinking into the soft soil. (206)

Leggy Lander

The Viking lander took about a year to reach Mars. The lander tested soil and sent pictures back to Earth.

Landing shock absorber. (207)

Weather instruments. (208)

Antenna (209)

Satellite dish. (210)

The color-test card checks that the camera shows the correct colors. (211)

This *container is for soil samples.* (212)

Super Sunset

The Viking lander took thousands of pictures of Mars. Each picture took about ten minutes to build up, in sections, as the camera moved around slowly. (213)

Red Desert

Mars is very cold. It has many dead volcanoes, craters, and dried-up rivers. It looks like a rusty, rocky desert. (214)

JUPITER AND SATURN

These two giants are the largest planets in our solar system. Jupiter is made of liquid, so it is not solid enough to land on, but if you could drive a car around its equator, it would take you six months of nonstop traveling. A similar journey around the Earth's equator would take only two weeks. Saturn is a beautiful planet with shining rings around its middle. Both planets spin around very fast, pulling the clouds into stripes.

S A T U R N (215)

Spinning Saturn

Saturn is a giant, spinning ball of liquid held together by gravity. This photograph shows a band of clouds and the rings. (216)

Seven Cold Rings

Saturn's rings are made up of glittering pieces of ice like trillions of snowballs. (217)

Voyager Voyages

The Voyager project sent probes to explore Jupiter and Saturn. Voyager 1did its job so well that Voyager 2 was rerouted to go to Uranus and Neptune.

Television cameras (218)

Dish antenna (219)

Radio antenna (220)

A power supply is carried on the probe. It does not use solar power because it is working so far from the Sun. (221)

This disk has pictures of Earth and sounds, such as a baby crying and music. If aliens find the disk, it will tell them about Earth. (222)

One-way Ticket

The journeys of Pioneer 10 and 11 and of Voyager 1 and 2 passed several of the planets. These spacecraft are now heading for the stars. (223)

JUPITER

(224)

Red Storm

Jupiter, like Saturn, is a huge ball of liquid. It has icy clouds and a giant red spot that is the center of a huge storm. (225)

Swirling winds blow Jupiter's clouds into a hurricane-like storm. (226)

Mega Moons

Jupiter has 16 moons circling around it. The largest is called Ganymede. It is bigger than Mercury. (227)

Pioneer's Pictures

The program for sending unmanned spacecraft to Jupiter was called Pioneer.

Pioneer 10 succeeded, so 11 went on to Saturn. Both sent back lots of pictures.

Pioneer (228)

Power supply (229)

Asteroid and meteor detector (230)

Sun sensor (231)

Dish antenna (232)

CONCLUSION

I hope this book has been an enjoyable introduction to earth and space.

To continue your study of earth and Space, please read:

Introduction to Earth and Space II.

Yours truly,

Danny Spinks

APPENDIX

DANNY SPINKS

AND FIANCE &
POET
CHUNG M NOAH

Love

Love is calling mom
Love is sitting on a rock,
Love is a basketball net,
Love is fish fillet,

— Chong Noah

CHILD

Is my child a flower,
Blooming subtly with delight.
Has my husband offered,
this sensation,
It was you my child,
made me to cry.

Chong Noah

Ground Hog Day
Man envys the Hair of Animals,
Venus is yellow.
　　　Color of Jealousy
God made man.
In the Image of Himself.
Therefore Venus is God,
Statue of A Beutiful
Woman, is Venus the Goddess.
Zeus is the God of the Earth.
Jesus, was the Son of the God

　　　　Chong Noah

www.ingramcontent.com/pod-product-compliance
Lightning Source LLC
Chambersburg PA
CBHW040827180526
45159CB00001B/92